The Strategic Threat of Inevitable Climate Change

At present, governments' attempts to limit greenhouse-gas emissions through carbon cap-and-trade schemes and to promote renewable and sustainable energy sources are probably too late to arrest the inevitable trend of global warming.

—Jasper Knight and Stephan Harrison[1]

The global climate is changing. Scientists around the world are working diligently to understand the causes driving the ongoing change and to what extent the climate will continue to change in the future. While they do so, leaders across the globe, both in government and the private sector are confronted with the implications of an uncertain future in which a changing climate may have significant impacts on nearly every aspect of our lives. While scientific projections of future climate conditions can vary considerably, the impacts on the climate already experienced as a result of global warming portend a future characterized by distinct winners and losers, including governments, industries, economies, societies and, at the very extreme, civilization and the world order as it exists today. It is imperative then that leaders at all levels determine how best to address the challenges of climate change. The actions and decisions leaders make in the face of the significant and far-reaching implications of climate change are vital to the continued security and prosperity of the United States.

While the terms climate change and global warming have become somewhat synonymous of late, particularly in popular media, political and economic discussions, they do not mean the same thing. As used most commonly in scientific discourse:

> Climate Change - A significant and persistent change in the mean state of the climate or its variability. Climate change occurs in response to changes in some aspect of Earth's environment: these include regular changes in Earth's orbit about the sun, re-arrangement of continents

through plate tectonic motions, or anthropogenic modification of the atmosphere.

Global Warming - The observed increase in average temperature near the Earth's surface and in the lowest layer of the atmosphere. In common usage, "global warming" often refers to the warming that has occurred as a result of increased emissions of greenhouse gases from human activities. Global warming is a type of climate change; it can also lead to other changes in climate conditions, such as changes in precipitation patterns.[2]

For the purposes of this paper global warming will mean the warming of the earth's surface a result of increased greenhouse gasses in the atmosphere. Climate change refers to the long term change in the measures of the climate including temperature, precipitation, wind, clouds and others factors.

Climate change is not a new phenomenon. The earth's climate has undergone significant changes, including wide variations in surface temperatures, precipitation patterns, sea levels and ice coverage throughout its history. The idea of global warming was first introduced in 1896 by Svante Arrhenius, a Swedish physical chemist. He estimated that doubling the level of carbon dioxide in the atmosphere would raise the mean global temperature by several degrees.[3] In 1980, ice core samples drilled in the Greenland and Antarctic ice caps, which allowed scientists to measure the level of carbon dioxide and temperature back to the last ice age, demonstrated a clear link between CO2 levels and temperature[4]. By the end of the 1980s, there was a general consensus among climate scientists that global warming was a serious risk and was attributable to the increase in greenhouse gasses in the atmosphere from the burning of fossil fuels[5].

In the ensuing two decades, much of the debate over climate change and how to respond has centered on its causality. One on side of the debate are those who maintain that human activity is causing, or considerably exacerbating global warming and its resultant impacts on regional and global climates. This is referred to anthropomorphic intervention in the climate[6]. On the other side are those who advocate that the current trend in global warming and climate change are part of the natural changes in the earth's climate that have occurred throughout history, caused by natural phenomenon, with minimal influence by human activity.[7]

As the debate over climate change in the US has centered on the extent to which human activity is contributing to global warming, so too has the proposed strategies to deal with the challenges climate change presents to the nation's security and prosperity. This paper posits that climate change is ongoing and continued change is inevitable. Climate change represents a significant threat to the security and prosperity of the United States. The current emphasis by the US government on efforts to mitigate anthropomorphic intervention in the climate will not have any appreciable effect on global warming and resulting climate change. The US should focus its efforts and resources on understanding and adapting to a future in which the climate is likely to be radically different than it is today.

Climate Change is Ongoing

There is credible evidence that the earth's climate is undergoing significant change. This is the conclusion drawn by the numerous governmental and independent agencies and organizations after comprehensive study of the extensive body of scientific evidence, research and modeling conducted and peer-reviewed by climate scientists over the past three decades. Prominent among those organizations and agencies are the United Nations' International Panel on Climate Change, the United States Global Change Research Program, the National Academy of Sciences from the

G8+5 countries, the US National Oceanic and Atmospheric Administration, and the United Kingdom's Government Office for Science. All agree that climate change is occurring and that human activity is responsible[8]. Their conclusions are summed up in the following statement from the National Academy of Sciences:

> A strong, credible body of scientific evidence shows that climate change is occurring, is caused largely by human activities, and poses significant risks for a broad range of human and natural systems.... Some scientific conclusions or theories have been so thoroughly examined and tested, and supported by so many independent observations and results, that their likelihood of subsequently being found to be wrong is vanishingly small. Such conclusions and theories are then regarded as settled facts. This is the case for the conclusions that the Earth system is warming and that much of this warming is very likely due to human activities.[9]

The US Global Change Research Program is charged with providing Congress and the President with an overarching and comprehensive assessment of the science of climate change.[10] Their report from 2010, *Global Climate Change Impacts in the United States*, states:

> Climate-related changes have already been observed globally and in the United States. These include increases in air and water temperatures, reduced frost days, increased frequency and intensity of heavy downpours, a rise in sea level, and reduced snow cover, glaciers, permafrost, and sea ice. A longer ice-free period on lakes and rivers, lengthening of the growing season, and increased water vapor in the atmosphere have also been observed. Over the past 30 years, temperatures have risen faster in winter than in any other season, with average winter temperatures in the Midwest and northern Great Plains increasing more than 7°F. Some of the changes have been faster than previous assessments had suggested.[11]

Most significantly, the surface temperature of the earth has risen .8 degrees Celsius (1.4 degrees F) since 1890[12]. While this may seem minor, it is central to ongoing climate change for three reasons. First, the rise in the earth's surface over the past century, and particularly since 1950, is part of an upward trend that continues to this day and is predicted to do so into the foreseeable future.[13] Globally, the sixteen warmest years on record have all occurred in the last twenty years while the last decade (2000-2009) was the warmest in the 160-year global record.[14]

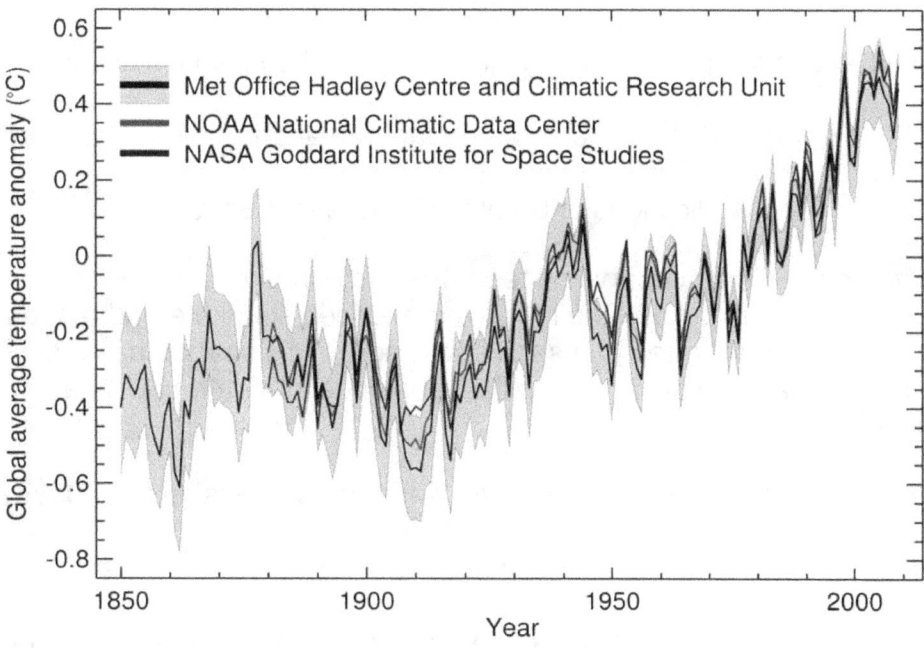

The three main records of global average surface temperature. Red line = NOAA record, blue line = NASA record, black line = Met Office/ UEA CRU record, with grey shading showing 95% confidence interval on Met Office/UEA CRU record. Source: Met Office Hadley Centre[15]

Figure 1. Global Average Surface Temperature

Second, while a study of warming and cooling trends across earth's history shows numerous warming periods, with temperatures rising at times above current levels, what is significant is the current rate of warming. According to the IPCC, "the largest temperature changes of the past million years are the glacial cycles, during which the global mean temperature changed by 4°C to 7°C between ice ages and warm interglacial periods."[16] Previous global warming in conjunction with ice ages occurred over much greater period of time, generally around 5000 years.[17] In contrast, on its current trajectory global temperatures are predicted to increase between 1.1 and 6.4

degrees Celsius (2-11.5 F) by the year 2100[18]. Global temperature rise equal to that of past ice age cycles will take place in less than a century. Ecosystems and species will be significantly challenged to adapt at that pace. Perhaps most importantly, this unprecedented rate of temperature increase underpins the scientific assertion that current global warming and climate change result from anthropomorphic interference.

Implications of Climate Change to the US

> Climate change is an observable fact. Regardless of cause, recent global temperature rise is outside the range experienced since the end of the last ice age approximately ten thousand years ago. This change has the potential to change many of the delicate balances that affect US national security.[19]

United Nations Secretary-General Ban Ki-moon, addressing the 2012 U.N. climate talks in Doha, Qatar described climate change as an "existential challenge for the whole human race".[20] While this may be an extreme view, it certainly portends that climate change could represent a significant threat to the security and prosperity of the United States. The idea that climate change would adversely affect the United States and its interests is not new. Climate change took on increasing importance for the US government over the past decade. In 2003, the Department of Defense conducted a study on the potential national security implications of abrupt climate change. The report presented a very dire future in which climate change leads to increased instability and conflict around the world over diminishing vital resources including food and water.[21] In 2008, Congress directed the Department of Defense to include potential impacts of climate change in all iterations of its Quadrennial Defense Review. The 2010 QDR stated that "while climate change alone does not cause conflict, it may act as an accelerant of instability or conflict, placing a burden to respond on civilian institutions and militaries around the world."[22] As recently as 2011, the Defense Science Board, in a study on the implications of climate change on security, stated "climate change is likely to have the greatest impact on security through its indirect effects on conflict and vulnerability" and that "climate change is more likely to be an exacerbating cause for failure to meet basic human needs and for social conflict, rather than the root cause."[23]

Much of the discussion regarding the implications of climate change has centered on the potential for increased conflict, violence and instability. While such scenarios do represent a threat to US national security interests, there is a more significant threat to the US presented by the changing climate. The source of US power, and thus its security and prosperity, is directly linked to the strength of the US economy. Globalization has inextricably linked national economies together, including that of the US. Small disruptions in world markets can have a cascading effect throughout the system. As an example, real or perceived threats to the oil supplies drive up the price of gasoline and petroleum products which ripples across economic sectors increasing costs in key industries including transportation, housing, energy, food, agriculture, government and military. Climate-driven increases in conflict, violence and instability around the world could significantly affect the US economy, undermining US strength and security.

Rising sea levels, increased temperatures and changing patterns of wind and precipitation are predicted to cause a significant increase in the frequency and severity of weather related disasters.[24] This will have a direct impact on the US economy. As an example, Hurricane Sandy caused an estimated $60 billion in damages to property and infrastructure and a sizable loss in economic activity. The extend climate change influenced Hurricane Sandy has not yet been determined. However, the scale and intensity of the storm may portend the costs and economic impacts in a future in which climate change makes storms, hurricanes, flooding drought and wildfires more frequent and severe.

Climate change, driven by global warming, may also undermine US economic strength as it stresses the US and the world's ability to support and sustain its population. For example, across the US almost 80 percent of agricultural land experienced drought in 2012, which made it more extensive than any drought since the 1950s.[25] Crop losses were roughly estimated to exceed $20 billion[26]. While the efficacy of the US agricultural sector, supported by US government subsidies, absorbed these losses with marginal disruptions to the US economy, sustained drought conditions in the central US could result in a return to the Dust Bowl conditions of the 1930s, with ensuing economic impacts for the US.[27]

Changing rainfall patterns resulting in reduced snowfall and widespread drought have brought portions of the Mississippi River to near record low levels, threatening commercial shipping on one of the nation's most critical economic waterways[28]. Along the US Atlantic coast from Cape Hatteras, North Carolina, to Boston, Massachusetts sea levels are increasing three to four times faster than rates of sea-level rise globally[29]. Sea levels on this stretch of coast have climbed by between 2 and 3.7 millimetres per year since 1980, whereas the global increase over the same period was 0.6–1.0 millimetres per year[30]. The Atlantic coastline contains numerous cities and ports vital to the economic vitality of the US, including Miami, Charleston, Norfolk, Boston, Philadelphia and New York. These cities are vulnerable to the impacts of climate change. Costs to mitigate their risk would be substantial, surpassed only by the costs of the damages they are likely to sustain from rising sea levels, increased storm surges and coastal flooding.

These and other impacts of the changing climate directly threaten the US economy and thus its security and prosperity. In addition, in its role of global leadership, the costs of global warming will manifest through increased requirements to provide humanitarian assistance, disaster relief and other types of assistance around the world. In addition to increased natural disasters, for nations absent the extensive resources of the US, the impacts of climate change will be far more damaging to their people, economies and political stability.[31] For example, sea levels have risen about 8 inches globally in the past century and are predicted to rise by another one to four feet in this century, increasing the risk of erosion, storm surge damage and flooding of coastal communities. As the US rebalances to Asia Pacific as a region vital to US economic, political and security interests, climate scientists predict that in the coming century:

> All coastal areas in Asia are facing an increasing range of stresses and shocks, the scale of which now poses a threat to the resilience of both human and environmental coastal systems, and are likely to be exacerbated by climate change. The projected future sea-level rise could inundate low lying areas, drown coastal marshes and wetlands, erode beaches, exacerbate flooding and increase the salinity of rivers, bays and aquifers. With higher sea level, coastal regions would also be subject to increased wind and flood damage due to storm surges associated with more intense tropical storms. In addition, warming would also have far reaching implications for marine ecosystems in Asia.[32]

Many countries and millions of people in the Asia Pacific and around the world will be significantly impacted by climate change. While dealing with the high costs of climate change at home, the US may not have the resources to act globally, but will certainly be expected to provide support and assistance to its key partners and allies as they cope with its impacts. US leadership and it resources will be essential for the current global economic and political system to adapt to an environment so drastically altered by climate change. Such leadership will incur substantial economic costs, leveraged against US economic strength, prosperity and security.

The Science of Inevitability

Human activity since the dawn of the Industrial Revolution has already led to a substantial increase in Greenhouse gasses in the atmosphere including carbon dioxide (CO_2), methane (CH_4), nitrous oxide (N_2O) and the halocarbons (a group of gases containing fluorine, chlorine and bromine)[33] According to the IPCC's report on climate change from 2007:

> Since the start of the industrial era (about 1750), the overall effect of human activities on climate has been a warming influence. The human impact on climate during this era greatly exceeds that due to known changes in natural processes, such as solar changes and volcanic eruptions[34]

Having accepted that global warming and climate change are being caused by human activity, scientists and world leaders have focused their response to climate change on efforts to limit Greenhouse gas. The United States led efforts during the 2009 UN Climate Change Conference that brought about the Copenhagen Accord. This nonbinding agreement, formally recognizing climate change as one of the greatest challenges of our time, agreed that

> To stabilize greenhouse gas concentration in the atmosphere at a level that would prevent dangerous anthropogenic interference with the climate system, we shall, recognizing the scientific view that the increase in global temperature should be below 2 degrees Celsius.[35]

This agreement to limit global warming to 2 degrees Celsius is at the center of US and international response to global warming and climate change. However, current scientific understanding and modeling project global warming beyond 2 degree Celsius. Additionally, there are political, economic and social barriers that preclude meaningful action to address global warming and climate change in the foreseeable future. The net effect is that the focus of the US and international strategy to address climate change, limiting global warming to 2 degree C, is likely to fail.

There is mounting evidence that the climate has already reached the point where human activity and climate feedback make continued climate change inevitable, regardless of any human intervention[36]. Some greenhouse gases are long-lived, meaning that once emitted their impact on surface temperature, precipitation, and sea levels are largely irreversible for more than 1000 years after emissions cease[37]. Sea level rise will continue as well, as changes in ocean heat content along with melting and dynamic ice loss in the Antarctic and Greenland will continue for centuries.[38] The climate, as a system, is affected by feedback, which scientists and modeling are trying to better understand and predict. As an example, increased heat radiated back from the higher concentrations of CO_2 is increasing ocean temperatures, causing sea levels to rise. Warmer temperatures and rising sea levels are causing increased melting of polar ice caps and the Greenland Ice Sheet. Increased ice melt will further exacerbate sea level rise. Additionally, snow and ice reflect a considerable amount of solar radiation back. Ice melt and reduced snowfall from rising temperatures and changes in precipitation patterns will force the earth to absorb more heat as less is reflected away[39]. As emissions of CO_2 are absorbed by the oceans, acidity levels increase which reduces the amount of CO_2 the ocean can absorb, resulting in greater concentrations of CO_2 in the atmosphere, reflecting more heat, warming the earth's surface further.[40] Another example, discussed later, is melting permafrost, which releases trapped CO2 and methane, contributing to increased CO2 in the atmosphere, increasing temperatures and further permafrost melting.[41]

The latest scientific observations and predictions are dire, with increasing evidence that global warming and its resultant climate change is occurring both faster and at a greater extent than predicted even a few years ago.[42] As discussed earlier, global temperatures have already increased nearly .8 degrees Celsius (1.4F) over the last 100 years, with more than 80% of this increase occurring since 1980.[43] That is nearly half way to the 2 degree C limit. Given the long-lasting effect of GHGs already in the atmosphere and climate feedback, global temperatures are likely to exceed the 2 degree limit even if GHG emissions were cut to zero today.[44]

Current climate models predict that if greenhouse gas emissions continue at their current rate the average global temperature will rise by an additional 2-6 degrees Celsius this century, 20 times the rate of warming recorded over the past 2 million years.[45] However, rather than remaining constant, the rate of greenhouse gas emissions is steadily increasing. A report from the United Nations Environment Programme (UNEP), published in December 2012, states that greenhouse gas emissions in 2012 reached 50 gigatons of carbon equivalent, 20% more than in 2000 and nearly 14% above where emissions need to be in 2020 to meet the 2 degree Celsius target. The report estimates that if not cut, emissions will reach 58 gigatons in 2020, 14 gigatons more than acceptable.[46] This is equal to the total emissions today of America, Europe and Russia combined.[47] Some developed countries, including the US and the European Union have reduced GHG emissions in recent years.

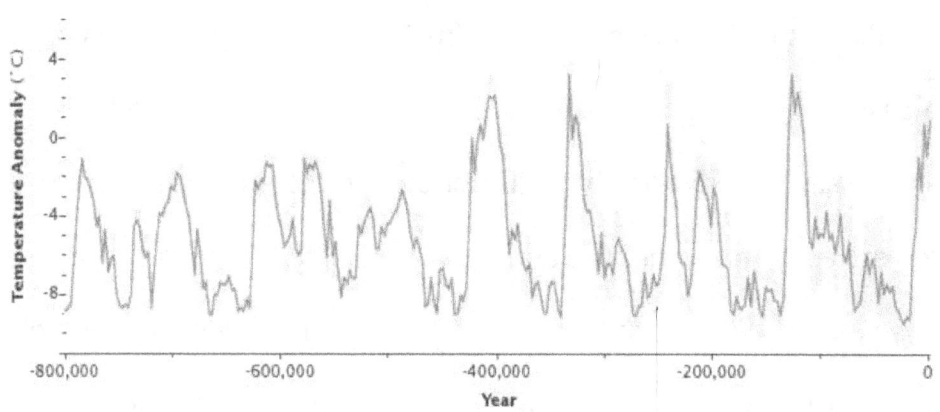

Scientists have built a record of Earth's past climates, or "paleoclimates." The paleoclimate record combined with global models shows past ice ages as well as periods even warmer than today. But the paleoclimate record also reveals that the current climatic warming is occurring much more rapidly than past warming events.

As the Earth moved out of ice ages over the past million years, the global temperature rose a total of 4 to 7 degrees Celsius over about 5,000 years. In the past century alone, the temperature has climbed 0.7 degrees Celsius, roughly ten times faster than the average rate of ice-age-recovery warming.[48]

Figure 2:

However, other countries, such as China, have significantly increased emissions, far outstripping reductions.[49] The lack of a concerted global effort to reduce greenhouse gas emissions it is almost certain that global warming will exceed 2 degrees Celsius this century. Indeed, the International Energy Agency (IEA) stated recently the chance of keeping global temperature rise at 2 degrees is virtually zero.[50]

As alarming as the above numbers are, they do not describe the extent of the challenge ahead. The effects of some climate feedback have yet to kick in, and in some cases these impacts are not incorporated in current climate modeling. Most significantly, the thawing of permafrost across much of the arctic region will contribute significantly to the increase in greenhouse gasses in the atmosphere. While climate projections predict a substantial loss of permafrost by 2100, the emissions of greenhouse gasses, including methane and CO2 from thawing permafrost, are not included into emissions predictions by UNEP or other climate models.[51] Permafrost contains large stores of frozen organic matter. As permafrost warms the organics in the soil decompose, producing methane and carbon dioxide that bubble up through the soil and enter the atmosphere.[52] A recent UNEP report estimates that warming permafrost could emit 43 to 135 gigatons of carbon dioxide equivalent by 2100 and 245 to 415 gigatonnes by 2200.[53] Such emissions could begin over the next few decades and continue for several centuries. Additionally, evidence now shows the Arctic is warming at twice the global rate. As permafrost melts, it amplifies global warming, creating, in effect, a permafrost carbon feedback loop, which may further exacerbate global warming.[54]

Evidence of the rate of warming in the Arctic, and in fact the most prominent example of global warming overall, is the extensive surface melting on the Greenland Ice Sheet and the significant reduction in summer sea ice in the Arctic. In its report to UN climate talks in Doha, the World Meteorological Organization (WMO) concluded that ice melt reached a "new record low" in the area around the North Pole. It estimated that the ice melt from March to September was 11.83 million square kilometers, the largest seasonal ice extent loss in the 34-year satellite record.[55] While the UMO report detailed other record-breaking weather events in 2012, including storms, flooding and drought, ice melt dominated the UNEP report. Ice melt, thus, is a crucial element of global warming not just for its environmental impact and contribution to climate change. It also represents the "public face" of global warming and climate change in a way that no other scientific data or localized climate change impact is able. However, despite presenting global warming as a real and significant threat, the ongoing loss of ice from the Arctic and the Greenland Ice Sheet has failed to spur global leaders to action.

Barriers Contributing to Inevitability

The international response to global warming and climate change, including the US, is coordinated by the United Nations. Nearly two hundred nations participate in the Convention of Parties to the United Nations Climate Change conferences. The United Nations position on global warming and climate change is that it is caused in large part by the emission of greenhouse gasses as a result of human activity.[56] However, while most political leaders and scientists now accept that emissions of greenhouse gasses are contributing to global warming, debate continues over what must be done to reduce such emissions, who must do so and, most importantly, when action must be taken.

In 2012, Doha, Qatar hosted the 18th annual UN climate conference. On the surface, it seems like a concerted international effort is being made to address climate change, with 192 nations participating. Indeed, the Doha conference, the first hosted in the Middle East, included approximately 9,000 participants, including 4,356 government officials, 3,956 representatives of UN bodies and agencies, intergovernmental organizations, civil society organizations and 683 members of the media.[57] Yet two decades worth of conferences has yielded few meaningful commitments to limit "dangerous anthropomorphic interference" with the climate.[58]

The notable exception is the Kyoto Protocol, agreed to by delegates at the 3rd COP in Kyoto, Japan in 1997. It committed industrialized countries and economies in transition to reducing their emissions of 6 greenhouse gasses by an average of 5% below 1990 levels during the first commitment period from 2008 to 2012, with percentage of reduction varying by country.[59] For example, the European Union agreed to reduce greenhouse emissions 20% by 2020 and is on track to meet or exceed those commitments.[60] While commendable, the EU's efforts are the exception rather than the rule among leading industrial nations.

Despite the Kyoto Protocols, global emissions of greenhouse gasses continue to rise. After reaching record high levels in 2011, emissions rose an additional 2.6% in 2012 to 35.6 billion tonnes, 58% above 1990 levels, the baseline year for the Kyoto Protocol.[61] While emissions in the EU and the US decreased in 2011, by 2.8 and 1.8 respectively, they increased in the developing world, with China emissions growing by 9.9% and India by 7.5%.[62] China is now the largest contributor to greenhouse gas emissions at 28%, with the US at 16%, the EU at 11% and India at 7%.[63]

Rather than adopt a more comprehensive and effective approach to climate change, the Doha conference simply extended the largely ineffective Kyoto Protocols until 2020. This continued lack of meaningful commitment to mitigate global warming is indicative of the stilted and ineffective approach to address climate change. More importantly, it virtually guarantees more years of inaction, leaving global warming unchecked in its rise to 2 degrees C of warming and beyond.

There are a number of reasons why nations refuse to make the significant changes necessary to reduce greenhouse emissions to levels consistent with limiting global warming to 2 degrees. First and foremost, the world's economy is depends on upon fossil fuels. The US, for example, consumed 6.87 billion barrels of refined petroleum products and biofuels in 2011[64] and used nearly a billion short tons of coal in producing 42% of the 4 trillion kilowatt hours of electricity generated in the United States that year.[65] China, for their part, surpassed the US as the world's biggest consumer in 2010 following its decades long burst of economic growth.[66]

As China's economy continues to expand, its demands for energy will increase, particularly for oil and coal. This is true for many nations as they focus on development and economic growth, particularly in the Pacific. According to the World Resources Institute, there are currently plans or construction ongoing for 1,199 new coal-fired power plants in the world spread across 59 countries, with China and India accounting for 76 percent of new plants combined.[67] The continuing rise in the demand for fossil fuels is certain to exacerbate global warming, pushing the world ever-closer to the point of inevitable and dramatic climate change.

To better frame the discussion on global warming, scientists have developed the idea of a global "carbon budget". This budget is an estimate of the maximum amount of carbon dioxide humans can emit into the atmosphere through mid-century and still hope to limit warming to the 2 degree target is 565 gigatons.[68] As discussed, meager international efforts to reduce greenhouse emissions through the increased use of renewable energy and improved efficiency have failed to arrest the overall rise in emissions across the globe, much less reduce them to the agreed upon 1990 levels. Climate models predict emissions will continue to rise roughly 3% a year. If that proves true, the 565 gigaton limit will be reached in only 16 years.[69]

Efforts to replace fossil fuels with renewable energy sources have been largely ineffective. Today in the US, after concerted efforts supported by sizable government investment in recent years, energy from renewable sources (hydropower, biomass, biofuels, wind, geothermal, and solar) in 2011 accounted for about 9.3% of total U.S. energy consumption and 12.7% of electricity generation.[70] This is consistent with the global average of approximately 10% of world energy consumption attributed to renewable sources.

While some progress has been made to increase the use of renewable energy sources, the EIA forecasts that by 2035, consumption of renewable energy will only rise to about 14% of total world energy consumption.[71] Thus, at current rates, seven years after the 565 gigaton carbon budget is reached, the world will still be reliant upon fossil fuels for 85% of its energy needs. Only Germany has undertaken a concerted effort, producing 25% of its power from renewable energy in 2012. Indeed, Germany has enacted legislation that renewable energy shall account for 35 percent of the electricity production by 2020, 50 percent by 2030, 65 percent by 2040 and 80 percent by 2050.[72] At issue is whether other leading producers of greenhouse gasses, including the US and China would follow suit.

There is a lack of political will to undertake such a costly transition given the integral role fossil fuels play in the economies of developed and developing nations.

Even political leaders who understand the dire future that likely lies ahead without significant reduction in CO2 emissions realize that the demands of today outweigh the potential dangers. Even absent outside influences and pressures, it is unreasonable to expect a political leader to commit their nation to significant hardships and sacrifice today to avert a dire future forecast to be decades or more away.

Additionally, there is considerable economic incentive to continue the reliance on fossil fuels. According to the Carbon Tracker Initiative, a team of London financial analysts and environmentalists who published a report to educate and inform investors about risks from climate change, leading fossil-fuel companies, and countries that act like fossil-fuel companies such as Kuwait and Venezuela, have proven coal, oil and gas reserves estimated to be 2,795 gigatons of carbon.[73] This is significant in that it is 5 times the 565 gigatons carbon budget.

Thus, even without any additional discovery, these companies have enough proven reserves to exceed the 2 degree temperature increase limit. Put another way, to stay under the 565 gigaton limit would require these companies, and the countries that stand to benefit financially, to keep 80 percent of known reserves in the ground.[74] That equates to more than $20 trillion dollars in assets not brought to market. Certainly, these companies are and will continue to invest heavily in ensuring any efforts to address climate change do not prevent them from exploiting these reserves. Meanwhile, these companies continue exploration to identify additional deposits of fossil fuels. They are also developing technologies that exploit known deposits from tar sands, deep seas and, ironically, in the Arctic, which is being opened by global warming.

Nations, even those committed to combating climate change, are stuck in the realities of the modern world and its dependence on fossil fuels. Secretary of State Hillary Clinton, in June of 2012, visited the Arctic on a Norwegian research trawler to see the impact of climate change. She stated "many of the predictions about warming in the Arctic are being surpassed by the actual data", describing the sight as "sobering".[75] Yet, the purpose of her visit was to discuss with other Arctic bordering nations how to divvy up the estimated $9 trillion in oil (90 billion barrels, 37 gigatons of carbon) that are becoming accessible as the ice melts.[76] Canada, long a stalwart of environmental advocacy, recently withdrew from the Kyoto treaty as the rising price of oil made the tar sands of Alberta, which contain an estimated 240 gigatons of carbon, economically viable.

A final barrier to action is the continuing debate over the realities of climate change and its causes. Climate scientists agree that climate change is ongoing and that its principal cause is human activity, specifically the emissions of greenhouse gasses from fossil fuels.[77] Ralph Cicerone, President of the National Academy of Sciences, summed up how extensively the scientific community has studied global warming to reach near consensus on the climate change and the impact human activity is playing:

> I think we understand the mechanisms of CO2 and climate better than we do of what causes lung cancer...In fact, it is fair to say that global warming may be the most carefully and fully studied scientific topic in human history.[78]

Yet dissent remains, undermining efforts of scientists and political leaders to garner support for meaningful action. Organizations like the George C. Marshall Institute, a think tank in Washington D.C. and others, including the Cato Institute, the Heritage

Foundation, the American Enterprise Institute, the Heartland Institute and the Competitive Enterprise Institute attempt to counter claims of climate scientists.[79] It is important to note that their efforts do not offer a scientifically tested alternative to the evidence of global warming and climate change. They focus on creating doubt as to the validity of the science, or to trustworthiness of climate scientists themselves. Rather than provide research and articles for peer-review and publication in scientific journals, these organizations focus on engaging through popular media, broadcast news interviews, editorials, articles in non-science magazines, blogs, etc. Their efforts are working. Opinion polls in recent years show many Americans disbelieve the scientific conclusions on climate change and anthropomorphic causes.[80]

Climate science today has a communications problem. Their consensus that scientific evidence clearly indicates that global warming and climate change are ongoing and caused by human activity has failed to gain popular support. They are not losing a debate over the science, but rather the information campaign for the version of climate change that will drive popular and political will. Scientists must do a number of things, and soon, if they hope to shape the discussion in the favor of their scientific finding and modeling. First, they need to engage in forums outside the scientific journals and conferences. Much of what climate scientists write is technical and found only in scientific journals. This is valuable for their peers, but of limited value to decision makers and the general public. Second, they need to reduce their equivocation over their evidence and predictions for climate change. Because science deals in probabilities and climate change is complex, scientists use terms like "likely" or "highly likely". These distinctions are important as scientific terms, but convey little to a broader audience not versed in the distinctions. Third, climate scientists need to educate the media and public on the difference in bona fides between them and leading climate skeptics. For example, Frederick Seitz, founder of the Marshall Institute, a prominent climate skeptic, is a solid-state physicist.[81] While a qualified and respected scientist, his expertise and experience is decidedly not related to climate science. Indeed, there are very few scientists in climate related fields who dispute anthropomorphic climate change.

Absent a concerted and effective communications campaign by climate scientists, popular and political support for decisive action to mitigate climate change is unlikely. The situation will have to first reach a point where climate change directly and unequivocally affects enough people that action, regardless of costs, is deemed essential. Given the trend scientific evidence and modeling shows the climate is currently on, it is certain that by the time its effects are so widespread and unequivocal to garner broad public and political consensus to act it will likely be too late to avert profound and enduring changes to the earth's climate and life as it exists today.

Conclusion

While scientists do not fully understand the climate in all its complexity, there is scientific consensus that ongoing global warming is in significant part a result of the increase in greenhouse gases in the earth's atmosphere from the consumption of fossil fuels. Climate scientists have warned of the dangerous impacts that a warming planet will have on the world's climate and the balance of ecosystems. The international response to the climate change has been muted, resulting in the current situation where fossil fuel emissions rise to new record levels every year.

The globe has already warmed nearly halfway to the internationally agreed upon limit of 2 degrees Celsius of warming if climate catastrophe is to be avoided. Given impacts of global warming at a mere .8 degrees, many scientists now believe that even 2 degrees of warming may be too much. Yet the demand for fossil fuels continues to increase as nations require more energy to foster economic growth and keep pace with the demands of burgeoning populations. Fossil fuel companies already possess enough reserves of coal, oil and gas to guarantee carbon emissions five times the amount scientists believe is the maximum the atmosphere can hold and still avert warming beyond the 2 degree limit. Advances in technology and political exigencies will enable the discovery and exploitation of deposits in areas currently restricted or unprofitable. Ironically, even the melting of Arctic ice by global warming may allow access to large reserves of fossil fuels. Economic and political realities will serve to continue the centrality of coal, oil and gas as the primary source of energy, transportation and indeed the global economy for decades to come.

Recommendations

What then, should the United States do in response to global warming? The US Government (USG) should accept that global warming cannot be limited to the 2 degree Celsius limit established by the Copenhagen Accords. To date, the US effort has been focused on finding politically and economically acceptable means of reducing greenhouse gas emissions to levels that would limit warming to 2 degrees. The USG should acknowledge that this is infeasible given the current climate trends and the political and economic incentives to delay meaningful cuts in carbon emissions.

The USG must change its strategy to address climate change from its current focus of limiting warming to two degrees to a strategy that focuses on adapting to a future environment reshaped by significant climate change. A target of six degrees Celsius by 2050 would be a reasonable assumption. The USG, in partnership with other nations, should commission a scientific panel to model the impacts of 6 degree C warming and its resultant effects across the spectrum of human activity, with an emphasis on food and water security, sea level rise, precipitation patterns, drought, flooding, human migration, regional national and global economies and political systems. Using these predictive models, the USG should take a whole of government approach to climate change, led by a senior US official at the cabinet level, responsible for planning and coordinating the US response to climate change. The focus of the US effort should be on preparing for and adapting to the challenges that climate change will present to US security and prosperity.

The DoD should include dramatic climate change scenarios, including a 6 degree C temperature increase and its concomitant effects in its assessment of future operational environments. The US, in concert with other industrial and developing nations, should provide incentives to both private and governmental investment in technologies that mitigate global warming, including Carbon Capture and Storage technologies and systems. The US along with the G8+5 countries, should consider imposing a "carbon tax" to hold the producers and consumers of fossil fuels financially responsible for the environmental impact resulting from the emission of CO_2. This is in line with environmental protections for industries in many nations. Revenues from such a tax could be used to offset the costs of climate change.

The world's climate is changing. The detrimental effects of that change are already evident around the globe. It will get worse, perhaps catastrophically so. Global warming beyond the limits scientists believe the delicate balance that makes up the world's ecosystem can absorb is either already inevitable, or will be in the very near future. Certainly it will be inevitable by the time world leaders are willing to make the drastic changes in the use of fossil fuels averting a climate catastrophe would require. The unprecedented rate of global warming and its resultant climate change has the potential to threaten the political, economic and social foundations of the world. The United States, to ensure its peace, prosperity, the welfare of its citizens and to preserve its global leadership role must prepare now for a future that will be dramatically reshaped by the inevitable climate change ahead.

The current US approach to climate change, and much of the world, with its emphasis on reducing and mitigating the anthropomorphic effects[82] on the climate, cannot and will not solve the problem. While the US and other nations should not abandon efforts to reduce carbon emissions and other practices that contribute to global warming and thus climate change, it is imperative the US focus its efforts on adapting to inevitable climate change.

Endnotes

[1] Jasper Knight and Stephan Harrison, "The Impacts of Climate Change on Terrestrial Earth Information Surface Systems," *Nature Climate Change*, 14 October 2012, http://www.nature.com/nclimate/journal/vaop/ncurrent/full/nclimate1660.html#/author-information (accessed 02 December 4, 2012).
[2] National Oceanic and Atmospheric Administration, "Climate Literacy," http://oceanservice.noaa.gov/education/literacy/climate_literacy.pdf (accessed February 6, 2012).
[3] Spencer Weart, "Global Warming: How Skepticism Became Denial," *Bulletin of the Atomic Scientists*, vol. 67, no.1 (January 2011): 42.
[4] Ibid., 44.
[5] Ibid.
[6] Anthropomorphic effects on climate change are those caused by human action. Over the past century, human activities have released large amounts of carbon dioxide and other greenhouse gases into the atmosphere. The majority of greenhouse gases come from burning fossil fuels to produce energy, although deforestation, industrial processes, and some agricultural practices

also emit gases into the atmosphere. For more see US Environmental Protection Agency website, http://www.epa.gov/climatechange/basics/ (accessed February 3 2013)

[7] Skepticism of anthropomorphic influence on the climate date back to the earliest days of global warming when initially proposed by Svante Arrhenius in 1896. The belief is that the earth's vast climate systems of atmosphere, ocean, rock and ice is self-regulating, maintaining its temperature and chemical composition over millennia. Human activity is thus too inconsequential to impact the durable "balance of nature". This ideal remains a central idea posited by those who doubt current scientific understanding of global warming and climate change. For more see: Weart, "Global Warming: How Skepticism Became Denial," 42-44

[8] Each of the organizations listed have published papers, studies or programs that state explicitly that global warming and climate change are occurring as a result of anthropomorphic interference. As an example, the Intergovernmental Panel on Climate Change, in its 4th Assessment Report, Climate Change 2007, states "Warming of the climate system is unequivocal, as is now evident from observations of increases in global average air and ocean temperatures, widespread melting of snow and ice and rising global average sea level" and that "Most of the observed increase in global average temperatures since the mid-20th century is very likely due to the observed increase in anthropogenic GHG concentrations," for more see Intergovernmental Panel on Climate Change, "Climate Change 2007: Synthesis Report," Cambridge University Press, New York, NY, 2007, 1-3.

[9] National Academy of Sciences, "Advancing the Science of Climate Change," National Academies Press, Washington D.C., 2010, 20-21.

[10] The U.S. Global Change Research Program (USGCRP) coordinates and integrates federal research on changes in the global environment and their implications for society. The USGCRP began as a presidential initiative in 1989 and was mandated by Congress in the Global Change Research Act of 1990 (P.L. 101-606), which called for "a comprehensive and integrated United States research program which will assist the Nation and the world to understand, assess, predict, and respond to human-induced and natural processes of global change," for more see http://www.globalchange.gov/about (accessed February 9 2013)

[11] Thomas R. Karl, Jerry M. Melillo, and Thomas C. Peterson, (eds.), "Global Climate Change Impacts in the United States," Cambridge University Press, 2009, 9.

[12] "Climate Change," *Bulletin of the American Meteorological Society*, November 2012, 1740. For a more detailed explanation of the significant rise in global temperatures in recent decades see Grant Foster1 and Stefan Rahmstorf, "Global Temperature Evolution 1979-2010," *Environmental Research Letters*, December 6, 2011 http://iopscience.iop.org/1748-9326/6/4/044022/

[13] National Academy of Sciences "Advancing the Science of Climate Change", 3

[14] United Kingdom Government Office for Science, "The World is Warming," http://www.bis.gov.uk/go-science/climatescience/world-is-warming (accessed February 6, 2013)

[15] Ibid.

[16] Intergovernmental Panel on Climate Change, "Climate Change 2007: The Physical Science Basis," Cambridge University Press, New York, NY, 2007, 465.

[17] Ibid.

[18] National Academy of Sciences "Advancing the Science of Climate Change", 40.

[19] Defense Science Board, "Trends and Implications of Climate Change for National and International Security," Washington D.C., October 2011, 62.

[20] Secretary-General Ban Ki-moon, "Remarks to UNFCCC COP18 High-Level Segment," Doha (Qatar), 04 December 2012 http://www.un.org/apps/news/infocus/sgspeeches/statments_full.asp?statID=1721 (accessed February 12, 2013)

[21] Peter Schwartz and Doug Randall, "An Abrupt Climate Change Scenario and Its Implications for United States National Security," October 2003, http://www.gbn.com/articles/pdfs/Abrupt%20Climate%20Change%20February%202004.pdf (accessed December 4, 2012) 14.

[22] Robert M. Gates, Quadrennial Defense Review (Washington, DC: U.S. Department of Defense, February 2010), 85.

[23] Defense Science Board Task Force, "Trends and Implications of Climate Change for National and International Security", xi.

[24] National Academy of Sciences "Advancing the Science of Climate Change, 2.

[25] United States Department of Agriculture Economic Research Service, "U.S. Drought 2012: Farm and Food Impacts", http://www.ers.usda.gov/topics/in-the-news/us-drought-2012-farm-and-food-impacts.aspx (accessed February 3, 2013) For a detailed analysis of US drought conditions throughout 2012 see National Climatic Data Center, "State of the Climate Drought Annual 2012", http://www.ncdc.noaa.gov/sotc/drought/ (accessed February 3, 2013)

[26] Jennifer Liberto, "Drought May Cost $20 Billion in Crop Insurance", *CNN Money*, August 3, 2012. http://money.cnn.com/2012/08/03/news/economy/drought-crop-insurance/index.htm (accessed February 6, 2013)

[27] Joseph Romm, "Desertification: The Next Dust Bowl", *Nature* 478, 27 October 2011, 450–451.

[28] United States Energy Information Administration, "Stretches of Upper Mississippi River Near Record Lows," January 7, 2013, http://www.eia.gov/todayinenergy/detail.cfm?id=9771 (accessed February 3, 2013). Large stretches of the Upper Mississippi have seen low water levels, although for the most part the Lower Mississippi River Basin (after the confluence of the Ohio and Mississippi rivers, near the Missouri, Illinois, and Kentucky borders) has remained at normal water levels. When levels start to drop below certain points, barge traffic is slowed because of congestion in the narrower portion of the river that remains navigable. The US Army Corps of Engineers has done extensive work in 2012 and into 2013 to ensure the river remains navigable to commercial traffic, particularly barges, through dredging and blasting of rocks from areas where water levels are dangerous low.

[29] Asbury H. Sallenger, Jr., Kara S. Doran & Peter A. Howd, "Hotspot of Accelerated Sea-Level Rise on the Atlantic Coast of North America," *Nature Climate Change*, June 24, 2012 http://www.nature.com/nclimate/journal/v2/n12/full/nclimate1597.html (accessed February 3, 2013)

[30] Ibid.

[31] Defense Science Board Task Force, "Trends and Implications of Climate Change for National and International Security," Washington D.C., October 2011, 53-55

[32] Intergovernmental Panel on Climate Change, "IPCC Fourth Assessment Report: Climate Change 2007, The Physical Science Basis" Cambridge University Press, New York, NY, 2007 https://www.ipcc.ch/publications_and_data/ar4/wg2/en/ch10s10-4-3.html#10-4-3-1 (accessed February 3 2013)

[33] Ibid., Chapter 1, 102.

[34] Ibid., Chapter 2, 135

[35] United Nations Framework Convention on Climate Change, "Report of the Conference of the Parties on its Fifteenth Session, Held in Copenhagen from 7 to 19 December 2009," 5, http://unfccc.int/resource/docs/2009/cop15/eng/11a01.pdf (accessed February 28, 2013)

[36] Defense Science Board, "Trends and Implications of Climate Change for National and International Security," xi.

[37] Susan Solomona, Gian-Kasper Plattner, Reto Knutti, and Pierre Friedlingstein, "Irreversible Climate Change Due to Carbon Dioxide Emissions," *Proceedings of the National Academy of Sciences of the United States*, Vol. 106, no 6., February 10, 2009, 1704-1709.

[38] Defense Science Board Task Force "Trends and Implications of Climate Change for National and International Security" 42.
[39] Thomas R. Karl, Jerry M. Melillo, and Thomas C. Peterson, (eds.), "Global Climate Change Impacts in the United States," Cambridge University Press, 2009, 17.
[40] National Academy of Sciences "Advancing the Science of Climate Change" 55.
[41] National Academy of Sciences "Advancing the Science of Climate Change" 56-57.
[42] "G8+5 Academies' joint statement: Climate Change and the Transformation of Energy Technologies for a Low Carbon Future," May 2009 http://www.nationalacademies.org/includes/G8+5energy-climate09.pdf (accessed February 6, 2013)
[43] NASA "How is Today's Warming Different than the Past" accessed December 9, 2012 at http://earthobservatory.nasa.gov/Features/GlobalWarming/page3.php
[44] John Carey, "Global Warming: Faster Than Expected?," Scientific American 307, no. 5 (October 2012) 50-55, in EBSCOhost (accessed February 8, 2013). 50-55.
[45] North American Space Agency, "How is Today's Warming Different than the Past" http://earthobservatory.nasa.gov/Features/GlobalWarming/page3.php (accessed December 9, 2012)
[46] United Nations Environment Programme, "The Emissions Gap Report 2012" November 2012 http://www.unep.org/publications/ebooks/emissionsgap2012 (accessed February 12, 2013) 1.
[47] "Theater of the Absurd" *The Economist,* December 1, 2012 http://www.economist.com/news/21567342-after-three-failures-years-un-climate-summit-has-only-modest-aims-theatre-absurd (accessed December 9, 2012)
[48] Ibid.
[49] "Record High for Global Carbon Emissions", Science Daily, December 4, 2012 http://www.sciencedaily.com/releases/2012/12/121202164059.htm (accessed December 15, 2012)
[50] "Theater of the Absurd" *The Economist*
[51] Kevin Schaefer et al., "Policy Implications of Warming Permafrost" United Nations Environment Programme, November 2012 http://www.unep.org/pdf/permafrost.pdf 17.
[52] Ibid., 18-19
[53] Ibid.
[54] Ibid., 18
[55] World Meteorological Organization "Provisional Statement on the State of Global Climate in 2012," November 28, 2012 http://www.wmo.int/pages/mediacentre/press_releases/documents/966_WMOstatement.pdf (accessed February 12, 2013)
[56] United Nations website "Gateway to the United Nations Systems Work on Climate Change," http://www.un.org/wcm/content/site/climatechange/pages/gateway/the-science (accessed February 10, 2013)
[57] "Summary of the Doha Climate Change Conference," *Earth Negotiations Bulletin*, International Institute for Sustainable Development, December 2012 http://www.iisd.ca/vol12/enb12567e.html (accessed December 15, 2012)
[58] Ibid.
[59] United Nations Framework for Climate Change, Kyoto Protocol webpage http://unfccc.int/kyoto_protocol/items/2830.php (accessed February 9, 2012)
[60] "Why the Doha Climate Conference Was a Success", *The Guardian*, http://www.guardian.co.uk/environment/2012/dec/14/doha-climate-conference-success (accessed December 15, 2012)
[61] "Record High for Global Carbon Emissions", *Science Daily*
[62] Ibid.

[63] Ibid.
[64] U.S. Energy Information Agency, http://www.eia.gov/tools/faqs/faq.cfm?id=33&t=6 (accessed December 15, 2012)
[65] U.S. Energy Information Agency, http://www.eia.gov/coal/ (accessed December 15, 2012)
[66] Spencer Swartz and Shai Oster, "China Tops US in Energy Use," *The Wall Street Journal*, July 18, 2010, http://online.wsj.com/article/SB10001424052748703720504575376712353150310.html (accessed December 15, 2012)
[67] Damian Carrington, "More Than 1,000 New Coal Plants Planned Worldwide," *Climate Central*, November 26, 2012 http://www.climatecentral.org/news/more-than-1000-new-coal-plants-planned-worldwide-15279 (accessed February 9, 2013)
[68] Bill McKibben, "Global Warming's Terrifying New Math", Rolling Stone July 19, 2012, accessed December 18, 2012 at http://www.rollingstone.com/politics/news/global-warmings-terrifying-new-math-20120719
[69] Ibid.
[70] U.S. Energy Information Agency, http://www.eia.gov/tools/faqs/faq.cfm?id=92&t=4 (accessed December 18, 2012)
[71] U.S. Energy Information Agency http://www.eia.gov/tools/faqs/faq.cfm?id=527&t=4 (accessed December 18, 2012)
[72] Stacey Leasca, "Germany Breaks Renewable Barrier" *Global Post*, July 30 2012, http://www.globalpost.com/dispatch/news/business/energy/120730/germany-breaks-renewable-energy-barrier (accessed December 18, 2012)
[73] Bill McKibben, "Global Warming's Terrifying New Math"
[74] Ibid.
[75] Ibid.
[76] Ibid.
[77] William R. L. Anderegg et al., "Expert Credibility in Climate Change" Proceedings of the National Academy of Sciences of the United States of America, June 21, 2010 http://www.pnas.org/content/early/2010/06/04/1003187107.full.pdf+html (accessed February 12, 2013)
[78] Ibid., 21.
[79] Naomi Oreskes and Erik M. Conwa, "Smoke, Mirrors and Climate Doubt," *Los Angeles Times*, Jun 08, 2010 http://search.proquest.com/docview/365582639?accountid=4444 A.13.
[80] Naomi Oreskes and Erik M. Conway, "Defeating the Merchants of Doubt", *Nature*, vol 465, June 2010, 686.
[81] Ibid.

www.ingramcontent.com/pod-product-compliance
Lightning Source LLC
Chambersburg PA
CBHW081819170526
45167CB00008B/3457